ISBN 978-1-333-36534-9
PIBN 10495615

This book is a reproduction of an important historical work. Forgotten Books uses
state-of-the-art technology to digitally reconstruct the work, preserving the original format
whilst repairing imperfections present in the aged copy. In rare cases, an imperfection in
the original, such as a blemish or missing page, may be replicated in our edition. We do,
however, repair the vast majority of imperfections successfully; any imperfections that
remain are intentionally left to preserve the state of such historical works.

1 MONTH OF
FREE
READING

at
www.ForgottenBooks.com

By purchasing this book you are eligible for one month membership to ForgottenBooks.com, giving you unlimited access to our entire collection of over 1,000,000 titles via our web site and mobile apps.

To claim your free month visit:

www.forgottenbooks.com/free495615

English
Français
Deutsche
Italiano
Español
Português

www.forgottenbooks.com

Mythology Photography **Fiction**
Fishing Christianity **Art** Cooking
Essays Buddhism Freemasonry
Medicine **Biology** Music **Ancient
Egypt** Evolution Carpentry Physics
Dance Geology **Mathematics** Fitness
Shakespeare **Folklore** Yoga Marketing
Confidence Immortality Biographies
Poetry **Psychology** Witchcraft
Electronics Chemistry History **Law**
Accounting **Philosophy** Anthropology
Alchemy Drama Quantum Mechanics
Atheism Sexual Health **Ancient History**
Entrepreneurship Languages Sport
Paleontology Needlework Islam
Metaphysics Investment Archaeology
Parenting Statistics Criminology
Motivational

SPECIERUM VARIETATUMQUE

GENERIS

DIANTHUS.

AUCTORE

FREDERICO   N.   WILLIAMS,   Soc. Linn. Sod.

———◆———

While in the course of their evolution plants have displayed progressive integrations, there have been at the same time progressive differentiations of the resulting aggregates, both as wholes and in their parts.—HERBERT SPENCER.

Vergebens dass ihr ringsum wissenschaftlich schweift,
Ein jeder lernt nur was er lernen kann.—GOETHE'S 'Faust.'

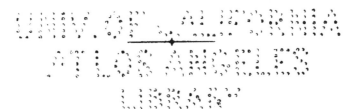
LONDON :

PRINTED BY WEST, NEWMAN AND CO., HATTON GARDEN, E.C.

——

1889.

SPECIERUM VARIETATUMQUE GENERIS *DIANTHUS:*

CUM DIAGNOSIBUS FORMARUM MINUS COGNITARUM.

———◆———

Tribus SILENEÆ.

Subtribus DIANTHEÆ.—Semina peltata, hilo faciali. **Embryo** rectus. Styli 2, a basi distincti.

1. VELEZIA.—Calyx anguste tubulosus, acute 5-dentatus, 5 vel 15-costatus. Stamina 5. Torus haud elongatus. Herbæ annuæ.

2. DIANTHUS. — Calyx multistriatus, 5-dentatus, bracteatus. Stamina 10. Torus sæpius in gynophorum plus minus elongatus. Herbæ sæpius perennes, rarius suffruticosæ.

3. TUNICA. — Calyx turbinatus vel elongato-tubulosus, obtuse 5-dentatus, 5 vel 15-costatus. Stamina 10. Torus parvus. Herbæ.

Genus DIANTHUS. — Calyx tubulosus, 5-dentatus, tenuiter et æqualiter multistriatus, nervis parallelis ad quodque sepalum 7, 9, vel 11 (3 in *Proliferastro*) parte membranacea inter 5 nervorum fasciculos; bracteis (*i. e.*, squamis calycinis) per paria calycem involucrum cingentibus, paribus inæqualibus. Petala 5; lamina abrupte attenuata sæpius in unguem elongatum, integra multidentata vel fimbriata, rarissime retusa, elegantia sæpe maculatâ. Stamina 10. Torus sæpius in gynophorum stipitiforme plus minus elongatus. Ovarium uniloculare; styli 2, a basi distincti. Capsula cylindrica oblonga vel rarius ovoidea apice dentibus valvisve quatuor dehiscens. Semina orbiculata vel discoidea, supra convexa, compressa, concavave parum infra, ad medium faciei interioris planæ vel concavæ umbilicata. Embryo rectus, in albumine sæpius excentricus. Herbæ perennes nonnunquam annuæ, rarius suffruticosæ, ramis articulatis, teretibus vel tetragonis, ad nodos tumescentibus. Folia exstipulata, angusta graminea, supremis subulatis, sæpe glauca; marginibus scabris. Inflorescentia terminalis; flores solitarii cymoso paniculati fasciculati vel capitati, vulgo rosei purpureive, nunc rubri, rare albi, nunquam lutei.

This genus is sufficiently marked off from *Velezia*, and is more nearly allied to *Tunica*. The generic description as given in the 'Genera Plantarum' is slightly modified to include Kunth's section *Kohlrauschia*, which has usually been included in *Tunica:* but, in the form and structure of the capsule and seeds, and the habit of the species, is nearer the type of *Dianthus.*

The genus is distributed through Europe, Temperate Asia, and N. and S. Africa.

Most of the forms are perennial; a few are annual or even biennial. The cæspitose habit of many of the perennial forms is due to the development of dense and leafy barren shoots, in which the internodes are almost suppressed. The rootstock produces barren shoots, and ascending flowering stems; the former, when not in the form of a rosette of leaves, are more or less decumbent from the crown, and curved away from the latter, so that the lower part of the stem is protected from fracture by the wind. The stems are either terete or angular, i. e., they may assume a cylindrical or prismatic form; in the latter case the number of the angles bears a definite relation to the phyllotaxis. As the leaves are opposite and decussate, stems that are not cylindrical have four angles. Angular stems are more frequently than not glabrous. In a few species the surface of the stem is viscid, but this viscidity is not confined to a glabrous surface. The barren shoots are more leafy than the ascending flowering stems, but they never become soboles or stolons. Throughout the genus the nodes are well-developed, and such as to give the stem a jointed appearance; and this is the more marked from the fact that the lamina springs direct from the stem without any intervening petiole. The internodes, which are almost suppressed in the barren shoots, in the flowering stems appear to bear some sort of relation to the leaves borne upon them, being sometimes equal in length to the leaves, sometimes double the length, and the ratio seems to obtain to the apex of the stem, where both internodes and leaves become shorter. Examples of shortened internodes are seen in the rosette of leaves at the base of the stem of D. Caryophyllus, in the fascicled leaves of the barren shoots of D. plumarius, and generically in the squamiform leaves beneath the floral organs. In the flowering stems of D. longicaulis the nodes are very distant. Since the intercalary growth is at its maximum, and persists longest at the base of each, in Dianthus similarly as it does in Grasses, it is at this point in the vicinity of the lower node of each internode, that is found localised heliotropic flexion. But certainly positive heliotropism is very feebly expressed in the genus. It is rarely that the stems spring as absolutely simple from the crown of the rootstock. They may affect a simplicity by a bifurcation deep in the cæspitose shoots at their origin, or by producing two divergent flowers at their termination. It may be difficult to determine to what extent their divergence from the rootstock is distinct, and how far the terminal capitulum may be reduced to a few flowers on a stem; perhaps a more trustworthy minor character is the acuteness of the angle of divergence near the base or apex. The characteristic mode of branching is cymose, frequently in dichotomies. It often happens in the upper part of the stem that the multiparous cyme, in impoverishing itself, continues as a biparous cyme, and that a biparous cyme degenerates in its turn to a uniparous cyme, developing only therefore one of its branches. Thus we see in some forms a biparous cyme terminating in helicoid or scorpioid uniparous cymes. The significance of the tumid nodes will be appreciated when their internal structure as distinct from that of the internodes is elsewhere discussed.

Leaves.—The two leaves of each whorl are opposite. The successive whorls of two leaves alternate so as to produce the decussate phyllotaxis. In *Syringa* and *Sambucus* the leaves are similarly arranged. It is the most frequent of alternate whorls of two leaves, but in *Rhamnus cathar-ticus* the two leaves of each whorl are usually at a slightly different level. The mature leaf is symmetrical, though in the aciculate leaves of a few species, such as *D. pinifolius,* this is not always patent: it is sometimes contorted at the base. Of the three parts of the typical leaf the petiole is absent; the absence of this is compensated for by the length of the sheath and the strength of the node. The midrib is usually strong, and deter-mines, by its position in the mesophyll and its relative size to that of the leaf, the character of the laminal surface, whether plane, carinate, or canaliculate. If it is placed nearer the inferior surface of the lamina, the leaf may be both carinate and canaliculate, or it may be carinate and plane, according to the thickness; but a leaf that is canaliculate on the superior surface is never plane on the inferior. The leaves are connate at the base, and frequently contracted at that point, resembling in this respect those of *Lonicera,* but never approach the perfoliate modification. The form of the mature leaf is linear and grass-like; those of the barren shoots are generally longer and broader than those of the flowering stems. The latter are more or less elongate-linear, while the former approximate the lanceolate-linear form; apices more or less acute, often very acuminate. The glaucescence characteristic of so many species temporarily disappears on pressure of the leaf between the moist fingers. The margin is entire, but is not sharp, as in that of the leaf of grasses, and is often ciliolate. The leaves of some species are of firmer texture than others, occasionally almost crassulaceous. They are firm and rigid, and conduplicate in vernation, and often remain triquetrous in æstivation, which is, however, more apparent than real, in consequence of the well-marked carination or prominent midrib. The cataphyllary leaf is represented in the well-marked leaf-sheath. This last partakes of the triplasic character of stipule, petiole, and scale. The uppermost leaves, distinct from the calyx, by some authors called bracts, are but squamiform modifications of the true foliage leaves. The true hypsophyllary leaves occur immediately beneath the floral organs, where they form an epicalyx of bracteiform scales, generally in two pairs.

Bracts.—These organs, above referred to, are very variable in form, texture, colour, and number. The number is not always constant in the same species, and where there are three pairs, the inferior pair are always longer and narrower than the other two pairs, thus approximating in form the uppermost leaves. The bracteiform appearance of the uppermost leaves, the variable form of the bracts themselves, the transformation of floral organs under cultivation, the gradual differentiation of the appen-dages of the caulomes, their character as accessories both of the nutritive and reproductive organs, demonstrate the truth of Goethe's law respecting the metamorphosis of similar organs of plants. The study of the bracts themselves as a link in the continuity of the nutritive and reproductive organs exhibits remarkable relations. Growing close under the perianth, they form part of the flower: they resemble the leaves in their decussate arrangement, and the petals in their general contour. In their colour they exhibit also transitional characters; sometimes herbaceous, like the foliage leaves, sometimes stramineous or coriaceous, sometimes tinted purple or red, like the petals and calyx-teeth. They are generally four in number to each flower, but vary from two to sixteen. When there is but a single pair, they rather approximate in form the foliage leaves. It also appears that one of their functions in many species is to protect the delicate tissue of the calyx from perforation by insects attempting to reach the nectar from below, instead of approaching from above the flower, and by this means brushing the anthers and fertilizing the ovary. Bracts with mucronate points are generally closely applied to the tube of the calyx, whilst those forms with acuminate points are patent. Groups of allied species may be thus distinguished. The length of the bract is determined rather from its aristate prolongation than from the length of the lamina. How closely the differentiation of the appendages of the essential reproductive organs is related to the form and variation of the nutritive organs is seen in those cases where groups of organs or of their appendages are transformed into those of next higher or lower type under changed conditions of environment. From an organogenetic point of view Willkomm seems inclined to regard them as an artificial rather than a natural factor in classification. It is true that in the atypical subgenera the uppermost leaves partake often of a bracteiform nature, instead of remaining herbaceous up to the floral heads, e. g., in *D. prolifer* the floral head is surrounded with an involucre of submembranous bracts, whilst each flower, except the central, is rolled up in coriaceous scales of a similar kind.

The Flower.—The morphology of the entire plant, and particularly of the flower, is that most favourable to entomophilous cross-fertilization. For in typical forms the barren shoots springing from the crown of the rootstock are short and decumbent, and thus expose the ascending and conspicuous flowering stems. And these are simple and paucifloral, or slightly branched, thus distributing the inflorescence over a proportionally wider area. The inflorescence is always terminal, and consequently not concealed by the dense barren shoots. The flowers are for the most part specious; when solitary, they are large and conspicuous; when small, they are collected into dense fascicles. The general coloration of the flower, and the frequent variegation of the lamina of the petal, is an inducement to the most fastidious of hymenopterous passers-by, associated as it often is with the most fragrant perfume, and easy accessibility. In the solitary or geminal flowers of *Caryophyllastrum* the petals are repand, large and specious, slightly excavated, and well supported on the cup of the strongly-ribbed calyx; in the fasciculate inflorescence of *Carthusian-astrum* and the capitular inflorescence of *Proliferastrum* their close apposition gives mutual support, and a large variegated area. The stamens and petals spring from an annular ridge of the torus encircling the gynophore, which latter is in the form of a prolonged axis between the calyx and the corolla. This ridge bears on its inner border a yellow, fleshy cushion, which is the nectary, and in its glands there is secreted the honey which attracts the fertilizing insects.

Perianth.—This consists of three parts—(1) the imbricated bracts; (2) the gamosepalous calyx; (3) the corolla of five distinct petals. The metamorphosis of the floral organs is well shown in the production of double flowers by cultivation. These double flowers in the carnations and pinks are produced by the assumption of a petaloid appearance by other organs. Not always, however, are the supernumerary petals either reverted stamens, or carpels, or petaloid sepals, but are instances of abnormal pleiotaxy, without reversion of essential organs to flower-leaves. The teratological tendency of the specific forms is towards petalody. The prolongation of the torus into a stipitiform gynophore seems to be analogous to the internode between the two successive whorls of foliage-leaves, and in this case separates the whorl of the calycine from the whorl of the corollary leaves.

Calyx.—Cylindrical, sometimes contracted above, occasionally as much as to become turbinate in form. It is multistriate, and the nerves are well-marked and parallel. The fasciculus of nerves to each lobe is bounded by a membranous part marking the division into sepals.

Petals.—Each of the two parts of the petal is quite distinct. The unguis is very pale; the lamina is bright-coloured, but frequently paler on the under surface. Chorisis in the petals is reduced to its simplest form, *viz.*, a capillary outgrowth at the junction of the lamina with the unguis, and is the homologue of the corona in *Silene* and the ligula in *Lychnis*. The segmentation of the lamina to such an extent as is seen in the fimbriate species is very unusual in flowers; and the retuse lamina of *D. glumaceus* is the least specialized modification of the laminar margin. In the same species petals may be entire and dentate, but they are never found in the same entire and fimbriate, or dentate and fimbriate.

Stamens.—The dichogamous development of the essential organs is usual throughout the genus, and thus functionally approaches the diclinous condition. In the individual flowers of the same plant the reproductive organs are almost invariably proterandrous, but this proterandry takes place at different times in the different flowers of the same plant; this is well seen in *D. deltoides*. This dichogamous mode of development undergoes a certain modification, to serve similar purposes in *D. Armeria* and other species. Though proterandrous functional activity is the form of dichogamy which generally obtains, it is not so well marked in this plant; consequently that modification obtains which is adapted in other cases to intercrossing among hermaphrodite flowers in which synanthesis is the rule, *i. e.*, where the maturation of the stigma is synchronous with that of the anthers. The petals are grooved by the respective filaments of the five outer stamens. The colour of the anthers is as variable as that of the laminæ of the petals; when the latter are dark, the anthers are often deep purple.

Ovary. — This organ is unilocular, with remains of apparent dissepiments. The styles are white and slender, and free at the base. The dehiscence of the capsule bears a more definite relation to the number of the styles and the phyllotaxis than to the other reproductive whorls.

Subgenus I. CARTHUSIANASTRUM. — Caudex annus v. perennis; perennibus turiones decumbentes steriles emittens atque multos caules adscendentes floriferos. Folia bracteiformia (suprema) sub floribus laxa. Inflorescentia cymoso-paniculata, v. fasciculis dichotomis v. capitulis aggregatis. Calyx subcylindricus. Petala semper dentata. Torus parum elongatus.

Sectio I. ARMERIUM.—Herbæ annuæ. Caules teretes. Bracteæ 2. Calyx dentibus 9–11 nerviis. Petala barbulata.

1. D. Armeria *L.* Nym. Consp.
   var. subhirsutus *Schur.*—Plus minus hirsutus. Flores majores.
   var. armeriastrum *Wolfn.*
   var. pseudarmeria *Wierzb.*
   var. subacaulis *Schur.*—Caules puberuli; flores atropurpurei;
       forma humilis.
2. D. pseudarmeria *M. B.* Nym. Consp.
3. D. corymbosus *Sib.* Nym. Consp.
   var. Poiretianus *Ser.*
4. D. tenuiflorus *Griseb.* Nym. Consp.
5. D. glutinosus *B. et Hld.* Nym. Consp.

Sectio II. SUFFRUTICOSI.—Perennes suffruticosi. Inflorescentia non densa; cymis paniculatis v. fasciculis dichotomis.

Subsect. 1. TUBULOSI.—Calyce apice non attenuato.

6. D. arboreus *L.* Nym. Consp.
7. D. fruticosus *L.* Nym. Consp.
8. D. pendulus *B. et Bl.* Boiss. Fl. Or.
9. D. actinopetalus *Fzl.* Nym. Consp.
10. D. elegans *d'Urv.* Nym. Consp.
11. D. Bisignani *Ten.* Nym. Consp.
12. D. virgatus *Pasq.* Nym. Consp.—Caules 60 centim., ramosi. Folia elongato-linearia, acuminata. Inflorescentia dichotome paniculata, rosea. Bracteæ 12, ovatæ, acutæ, adpressæ.
13. D. Bertolonii *Woods.* Nym. Consp.—Caules 60 centim., ramosi. Folia linearia, acuta, carnosa, carinata. Flores fastigiati. Bracteæ 8, lanceolatæ, acuminatæ ad ¼ calycis longitudinem.
14. D. juniperinus *Sm.* Nym. Consp.
15. D. aciphyllus *Sieb.* Nym. Consp.
16. D. rigidus *M. B.* Nym. Consp.

Subsect. 2. Contracti.—Calyce apice attenuato.

17. D. Friwaldskyanus *Boiss.* Nym. Consp.
18. D. gracilis *S. et S.* Nym. Consp.
    var. pumilus *Boiss.*
    var. armerioides *Griseb.*
19. D. biflorus *Griseb.* Nym. Consp.
20. D. Mercurii *Hldr.* Nym. Consp.

Sectio III. Carthusianum. — Herbæ perennes. Inflorescentia densa, capitata. Petala nonnunquam imberbia.

Subsect. 1. Microlepides. — Caules teretes. Bracteæ lanceolatæ. Calyx dentibus lanceolatis.

21. D. trifasciculatus *Kit.* Nym. Consp.
    var. heptaneurus *Griseb.*
22. D. transsylvanicus *Schur.* Nym. Consp.
    var. lancifolius *Tsch.* Nym. Consp.
23. D. nardiformis *Jka.* Nym. Consp. — Cæspitosus, 17 centim.
    Folia stricta, patentia, vaginâ caulis diam. triplo longiore.
    Flores cymoso-paniculati. Folia involucralia et bracteæ
    scariosa pallida. Bracteæ 4 obcordatæ. Calyx apice
    attenuato, dentibus acutis. Petala imberbia.
24. D. viscidus *B. et Ch.* Nym. Consp.
    var. olympiens *Boiss.*
    var. parnassicus *Boiss.*
    var. Grisebachii *Boiss.*
    var. alpinus *Boiss.*
25. D. tymphresteus *Hldr. et Sart.* Nym. Consp.
26. D. trifasciculatus *Schur.* Pl. Transsylv.
27. D. japonicus *Thunb.* Fl. Japonica, p. 183.
28. D. Muschianus *B. et R.* Boiss. Fl. Or.
    var. major *Boiss.*
29. D. erinaceus *Boiss.* Boiss. Fl. Or.
    var. Webbianus *Parol.*
30. D. pinifolius *S. et S.* Nym. Consp.
31. D. liburnicus *Bartl.* Nym. Consp.
    var. ligusticus *V.*
    var. propinquus *Schur. hb.*
32. D. cibrarius *Clem.* Boiss Fl. Or.
33. D. calocephalus *Boiss.* Boiss. Fl. Or.
34. D. giganteus *d'Urv.* Nym. Consp.
35. D. banaticus *Heuff.* Nym. Consp.
    var. biternatus *Schur.*
    var. pruinosus *Jka.*
    var. ponticus *Wahl.*

Subsect. 2. Carthusianoides. — Folia stricta. Bracteæ siccæ. Calyx dentibus lanceolatis. Petala obovato-cuneata, barbulata.

36. D. carthusianorum *L.* Nym. Consp.
    varr. prox. 'eu-carthusianorum.'
      a. eu-carthusianorum *L.*

b. nanus *Swt.*—Caules pauci 7 centim. Bracteæ flavæ, nunc brunneæ. Lamina $= \frac{1}{2}$ unguis, purpurea.

c. humilis *Brügg.* — 10 centim. Folia elongato-linearia, 10 mm. Lamina alba $= \frac{1}{2}$ unguis.

d. vaginatus *Vill.*

varr. prox. ' atrorubens *All.*'

e. carmelitarum *Reut.*

f. anisopodus *Ser. mss.*

g. Pontederæ *Kern.* — Folia radicalia linearia, caulinia elongato-linearia, vaginæ longitudine 4-plo longiore latitudine. Inflorescentia fasciculis 6–12 florum. Calyx dentibus triangularibus. Lamina $= \frac{1}{2}$ unguis, supra rubello-purpurea infrà rosea.

h. sanguineus *Vis.*

i. atrorubens *All.* — Folia elongato-linearia, vaginis purpureis caulis diam. duplo longioribus. Inflorescentia densa. Lamina $= \frac{1}{2}$ unguis, atrorubens.

j. pauciflorus *Brugg.*—38 centim. Fasciculis 2–6 florum. Lamina $= \frac{1}{2}$ unguis, atrorubens.

k. chloæphyllus *Schur.* — Glaucus 13 centim. Folia elongato-linearia ; vaginâ caulis diam. 3-plo longiore. Lamina rubra.

l. congestus *G. et G.*

m. roridus *Schur.*—Folia elongato-linearia. Inflorescentia capitulis 6–8 florum purpureorum. Lamina $= \frac{1}{2}$ unguis.

n. saxigenus *Schur.*—Caules 38 centim., numerosi, scabri. Folia elongato-linearia. Inflorescentia capitulis 2–6 florum purpureorum.

o. australis *Panc.* — Caules puberuli. Folia radicalia elongato-linearia, caulinia linearia. Inflorescentia fasciculis 6–12 florum purpureorum.

p. minor *Schur.* — 19 centim. Bracteæ flavæ scariosæ. Lamina purpurea.

q. pratensis *Ptck.* — 30 centim. Folia linearia. Bracteæ flavæ. Lamina purpurea.

varr. prox. ' consanguineus ' (bracteis oblongo-ovatis).

r. consanguineus *Schur.*

s. subfastigiatus *Schur.*—Caules 36 centim., suprà teretes. Folii vagina caulis diam. 4-plo longiore. Inflorescentia fastigiato-capitata. Bracteæ flavæ.

t. rupicolus *Schur.*

u. lancifolius *Schloss. et Vukot.*

varr. prox. ' pumilus.'

v. sabuletorum *Heuff.*

x. pumilus *Schur.*

varr. prox. 'ferrugineus.'—Bracteis diam. trans. æqualis apice basique, lateraliter per marginem, paralleli-planis.

y. gramineus *Schur.*

z. ferrugineus *L.*

a^2. tenuifolius *Schur.*—Caules tenues. Folia linearia. Inflorescentia capitulis 2–6 florum rubrorum.

varr. prox. ' atrorubens *Jacq.*'

b^2. ternatus *Schur.* — Caules 64 centim. Inflorescentia fasciculis ternatis. Bracteæ rubræ. Lamina = $\frac{1}{2}$ unguis.

c^2. surulis *mihi*. — Folia acicularia; vagina rubra, caulis diam. 4-plo longiore. Flores utroque capitulo 4–6. Lamina purpurea.—Transsylvania austr. Herb. Kew.

d^2. vaginatus *Ch.v.* cf. *d.*

e^2. giganteiformis *Borb.*

f^2. atrorubens *Jacq.* — Caules 30 centim. Folia linearia; vagina caulis diam. triplo longiore. Inflorescentia capitulis 3–6 florum atrorubentium.

varr. prox. ' vulturins.'

g^2. alpestris *Balb.* — Inflorescentia capitulis 4–6 florum atrorubentium.

h^2. glaucophyllus *Wierzb.*

i^2. subalpestris *Schleich exs.* — Inflorescentia capitata. Bracteæ flavæ. Lamina suprà purpurea, infrà flava.

j^2. parviflorus *Schur.*

k^2. vulturius *G. et T.*

37. D. Knappii *A. et K.* Nym. Consp.
    var. rosulatus *Borb.*—Folii vagina atque calyx purpurei.

38. D. ambiguus *Panc.* — Glaber. Caules simplices 4-angulares. Folia elongato-linearia, acuminata, plana 3-nervia, vagina caulis diam. 4-plo longiore, patentia. Inflorescentia capitulis densis 22–28 florum. Bracteæ 4 ad calycis apices mucronatæ, patentia. Calyx purpureus, dentibus acuminatis 9-nervis.—Serbia.

39. D. Schlosseri *mihi*. Hb. Mus. Brit. coll. Schlosser apud Janobor in Croatia.—Glaber. Caules 45 centim., teretes, simplices. Folia elongato-linearia, acuminata 3-nervia, adpressa; vagina caulis diam. duplo longiore; radicalia 100 mm., caulinia 70 mm. Flores capitati. Bracteæ 6, obovatæ ad $\frac{1}{3}$ calycis partem mucronatæ adpressæ. Calyx purpureus apice attenuato.

40. D. pelviformis *Heuff.* Kerner, Schedæ Austro-Hung.

41. D. cruentus *Griseb.* Nym. Consp.

42. D. Lydus *Boiss.* Boiss. Fl. Or.

43. D. mœsiacus *Panc.* Pl. Serbicæ rariores Dec. III.

44. D. lilacinus *B. et Hld.* Nym. Consp.

45. D. intermedius *Boiss.* Boiss. Fl. Or.

46. D. pseudobarbatus *Bess.* Nym. Consp.

47. D. barbatus *L.* Nym. Consp.
    var. latifolius *Ser.*
    var. aggregatus *Poir.*
    var. rariflorus *Schur.*—Inflorescentia non densa, rosea.

48. D. subbarbatus *Schur.* Pl. Transsylv.
    var. pedunculatus *Schur.*

49. D. Borbasii *Vandas*. Œsterr. Bot. Zeit. 1887.—Glaber. Caules
    simplices 45 centim., 4-angulares. Folia elongato-linearia,
    carinata trinervia, acuminata; vaginâ caulis diam. 6-plo
    longiore. Inflorescentia capitulis 2–7 florum, infrà bi-
    fasciculata. Bracteæ pallidæ marginaliter brunneæ, ellip-
    ticæ. Calycis dentes acuminati purpurei 11-nervii. Lamina
    purpurea, $= \frac{1}{2}$ unguis.
50. D. capitatus *DC.* Nym. Consp.
    varr. major et minor *Griseb.*
    var. Pancicianus *mihi.*—Var. innominata de Pancie. Forma
       glabra floribus purpureis. Serbia.

Subsect. 3. MACROLEPIDES.—Bracteæ 4 ovatæ patentes.

51. D. compactus *Kit.* Nym. Consp.
    var. prelucianus *mihi.*—Calyx viridis. Lamina pallidissima.
       Apud Preluci com. Naszod in Hungaria: coll. F. Porcius.
52. D. crassipes *Willk.* Nym. Consp.
53. D. Girardini *Lam.* Nym. Consp.
54. D. asperulus *B. et Huet.*
55. D. collinus *W. et K.* Nym. Consp.
    var. subpaniculatus *Schur.*
    var. imereticus *Rupr.*—Glaber. Inflorescentia fasciculis florum
       parvorum.
    var. sylvaticus *Hpe.*—Calyx fuscus dentibus acutis.
    var. alpestris *Balb.*
    var. geminiflorus *Lois.*
56. D. hymenolepis *Boiss.* Boiss. Fl. Or.
57. D. polymorphus *M. B.* Nym. Consp.
58. D. cinnabarinus *Spr.*
    var. Samaritani *Hldr.*
    var. biflorus *S. et S.*
59. D. Bitlisianus *Ky.* Boiss. Fl. Or.
60. D. pseudobarbatus *Schur.* Fl. Transsylv.—Glaber, 45 centim.
    Folia linearia acuta 5-nervia plana; radicalia 72 mm.,
    vagina caulis diam. 2-plo longiore, caulinia 50 mm.,
    vagina caulis diam. æquante. Inflorescentia fasciculis
    trichotomis 12 florum. Bracteæ mucronatæ ad calycis
    apices. Calyx dentibus lanceolatis acuminatis purpureis.
    Lamina obovato-cuneata, rosea.
61. D. toletanus *B. et R.* Nym. Consp.
62. D. anticarius *B. et R.* Nym. Consp.
63. D. stenopetalus *Griseb.* Nym. Consp.

Subgenus II. CARYOPHYLLASTRUM.—Caudex perennis, herbaceus,
breves turiones steriles decumbentes, numerosos foliososque, et
caules floriferos adscendentes, emittens. Foliis veris bracteiformibus
nullis. Flores solitarii vel geminati vel rarius cymis laxis. Calyx
cylindricus valde costatus præsertim superne. Petala dentata,
integra vel fimbriata. Torus elongatus in gynophorum stipiti-
lorme.

Sectio I. FIMBRIATUM.—Bracteæ 4–16. Petala fimbriata.

Subsect. 1. SCHISTOSTOLON.—Caules ramosi, glabii.

64. D. monspessulanus *L.* Nym. Consp.
   var. alpestris *Hpe. et Stcinb.*
   var. condensatus *Kit.*
   var. acuminatus *Tsh.*
65. D. marsicus *Ten.* Nym. Consp.
66. D. liliodorus *Panc.* Nym. Consp.—Caules 34 centim., teretes.
   Folia radicalia 34 mm., caulinia 24 mm., 5–7-nervia;
   vaginâ caulis diam. æquante. Flores albi odoratissimi.
   Calyx apice attenuato, dentibus lanceolatis, acuminatis,
   7–9-nerviis. Lamina obovato-spathulata, imberbis, $= \frac{1}{3}$
   unguis.
67. D. controversus *Gaud.* Fl. Helvetica.
68. D. Waldsteinii *Sternb.* Nym. Consp.
69. D. floribundus *Boiss.* Boiss. Fl. Or.
70. D. saxatilis *P.* Nym. Consp.
71. D. Sternbergii *Sieb.* Nym. Consp.
72. D. serrulatus *Desf.* Mumby in Enum. Pl. Algeriæ.
73. D. Tabrisianus *Bien.*
74. D. purpureus *mihi.* Herb. Kew. Mt. Hermon, Palestina;
   remissus a collegio Syriaco, 1879.—Caules tenues, 4-
   angulares. Folia elongato-linearia, adpressa acuminata;
   vaginâ caulis diam. æquante purpureâ. Bracteæ 4 ovatæ,
   acuminatæ dimidium calycem, adpressæ stramineæ. Calyx
   purpureus, dentibus lanceolatis, acuminatis 7–9-nerviis.
75. D. prostratus *Jacq.* Boiss. Fl. Or.
76. D. squarrosus *M. B.* Nym. Consp.
77. D. Zeyheri *Sond.* Flora Capensis, i.
78. D. Holtzeri *Winkl.* Gartenflora, 1883; Pl. Turkestaniæ.—
   Cæspitosus. Folia stricta, acuminata; radicalia sub-
   spathulata, caulinia lineari-lanceolata. Bracteæ 2 obovato-
   lanceolatæ, mucronulatæ ad dimidium calycem. Calyx
   dentibus lanceolatis, acuminatis. Lamina alba, maculata,
   barbulata.
   var. ebarbata *Winkl.*
   var. flaccida *Winkl.*
79. D. spiculifolius *Schur.* Nym. Consp.
80. D. acicularis *Fisch.* Nym. Consp.
81. D. plumosus *Spreng.* Ledebour, Fl. Rossica.
82. D. Kuschakewiczi *Reg. et Schmal.* Pl. nov. et minus cognit.,
   fasc. v. p. 28, 1877. — Folia linearia, acuta, flaccida,
   patentia, 3-nervia. Bracteæ 4 inf. oblongo-ellipticæ, sup.
   ovato-ellipticæ acuminatæ ad dimidium calycem, adpressæ.
   Calyx dentibus linearibus, acuminatis. Laminæ oblongæ,
   imberbes, non contiguæ.—Alatau Turkestaniâ.
83. D. stramineus *B. et Hldr.* Boiss. Fl. Or.
84. D. robustus *B. et K.* Boiss. Fl. Or.

85. D. oreadum *Hance.*   Ann. d. Sciences Nat. (Bot.) 1866, p. 207. Herb. Hance no. 1720.
86. D. sinaicus *Boiss.*   Boiss. Fl. Or.
87. D. valentinus *Willk.*   Nym. Consp.
88. D. polylepis *Bien.*   Boiss. Fl. Or.

Subsect. 2. CYCAXOSTOLON.—Caules simplices, teretes.

89. D. graminifolius *Presl.*   Nym. Consp.
90. D. erythrocoleus *Boiss.*   Boiss. Fl. Or.
91. D. atomarius *Boiss.*   Boiss. Fl. Or.
92. D. fallens *Timb.*   Nym. Consp.
93. D. noëanus *Boiss.*   Nym. Consp.
94. D. petræus *W. et K.*   Nym. Consp.
    var. pseudocæsius *Schur.*—Glaucus.   Petala contigua.
    var. bohemicus *Meyer.*—Folia linearia.   Flores bini.
95. D. plumarius *L.*   Nym. Consp.
    var. hortensis *Schrad.*
96. D. hungaricus *P. (Rchb.).*   Nym. Consp.
97. D. arenarius *L.*   Nym. Consp.
98. D. Serpæ *Hiern.*   Trans. Linn. Soc. 1881, p. 17.
99. D. fimbriatus *M. B.*   Boiss. Fl. Or.   (+ D. Broteri + D. scoparius + D. catalaunicus + D. macronyx *Fzl.*, etc.).
   a. Cæspitosi glauci.   Bracteæ 8.
    var. mutica *mihi* (= D. scoparius *Fzl.* in Ky. exs. Pers.).
    var. mucronulata *mihi.* — Bracteæ orbiculares, mucronulatæ. Calyx apice non attenuato.—Herb. Kew.   Hispania.
    var. Hookeri *mihi.* — Folia caulinia 72 mm., atroviridia. Bracteæ foliaceæ, ovatæ, acuminatæ. — Herb. Kew. Mts. Himalaya.   Fl. Brit. India, var. innom.
    var. sclerophyllus *Willk.*
    var. brachyphyllus *Willk.*
    var. leptophyllus *Willk.*
   b. Viridis non cæspitosus.   Bracteæ 12.
    var. orientalis *Sims.*
   c. Virides non cæspitosi.   Bracteæ 4.
    var. incertus *Jacqm.*
    var. obtusisquameus *Boiss.*
   d. Non cæspitosi.   Flores rosei.   Bracteæ 6-8.
    var. brachyodontus *B. et Huet.*
    var. stenocalyx *Boiss.*—Calyx angustus.   Lamina pallida.
    var. macronyx *Fzl.*
    var. brevifolius *Boiss.*
    var. pogonopetalus *B. et K.*
    var. angulatus *Royle.*—Caules infra 4-angular.
    var. macrophyllus *Willk.*—Flores majores.   Bracteæ 6.
100. D. gallicus *P.*   Nym. Consp.
    var. lusitanicus *Sjögr.*
101. D. macranthus *Boiss.*   Nym. Consp.

Subsect. 3. GONAXOSTOLON.—Caules simplices, tetragoni.

102. D. micropetalus *F. M.*—Flora Capensis, i.
    var. scaber *F. M.*
    var. glabratus *Sond.*
103. D. tener *Balb.* Nym. Consp.
104. D. serrulatus *Schloss.* Nym. Consp. (*Cf.* 72).
105. D. serotinus *W. et K.* Nym. Consp.
106. D. canescens *Koch.* Boiss. Fl. Or.
107. D. crinitus *Sm.* Boiss. Fl. Or.
    var. tomentellus *Boiss.*
    var. crossopetalus *Fzl.*
    var. pubescens *Boiss.*

Subsect. 4. MONERESTOLON. — Caulis unicns ramosus in multos cauliculos glabros. Folia patentia recurva. Petala barbulata, non contigua.

108, D. libanotis *Labill.* Boiss. Fl. Or.
109. D. superbus L. Nym. Consp.
    var. speciosus *Rchb.*
110. D. Wimmeri *Wich.* Nym. Consp.

Sectio II. BARBULATUM. — Flores solitarii v. cymis laxis, rosei v. purpurei. Petala dentata, barbulata.

Subsect. 1. LEPIDACRIBIA.—Bracteæ scariosæ, atting. $\frac{1}{4}$–$\frac{1}{3}$ calycis longitudinem.

111. D. lusitaniens *Brot.* Nym. Consp.
112. D. lusitanoides *mihi.*—Cæspitosus, glaber. Caules 48 centim., ramosi, teretes. Folia canaliculata, radicalia linearia acuta, caulinia elongato-linearia, acuminata stricta adpressa; vaginâ caulis diam. æquante. Bracteæ 4 obovato-lanceolatæ mucronatæ adpressæ. Calyx dentibus lanceolatis acuminatis purpureis 9-nerviis. Lamina obovata.— Herb. Kew. " Palestine Exploration Society ; east of the Jordan."
113. D. cæspitosus *Thunb.* Flora Capensis, i.
114. D. cæsius *Sm.* Nym. Consp.
    var. flaccidus *Fieb.*
    var. pulchellus *Rchb.*
115. D. polycladus *Boiss.* Fl. Or.
    var. breviberbis *Boiss. herb.*
116. D. multipunctatus *Ser.* Nym. Consp.
    var. micrantha *Boiss.*—Lamina rosea minuta.
    var. striatella *Boiss.*—Nervi secus tubum brevissimi.
    var. holosericea *mihi.* — Forma syriaca. Herb. Mus. Brit.
        indumento vere holosericeo.
    var. axilliflora *Fzl.* (forte abnormis).

var. subenervis *Boiss.*—Calycis dentes ipsi fere enerves tubo totaliter enervi.

var. pruinosus *Post.* Journ. Linn. Soc. vol. xxiv. p. 422.

117. D. Colensoi *mihi.* Herb. Kew; Natal, prope litus. — Caules glabri 60 centim., 4-angulares. Folia stricta adpressa; radicalia 52 mm., 9-nervia oblongo-lanceolata obtusa, caulinia 21 mm. 7-nervia lanceolata acuta plana, vaginâ caulis diam. æquante. Flores solitarii. Bracteæ 6 adpressæ. Calyx dentibus lanceolatis acuminatis. Petala non contigua, lamina alba obovata.

118. D. zonatus *Fzl.* Boiss. Fl. Or.

119. D. brachyanthus *Schur.* Pl. Transsylv.

120. D. viridescens *Vis.* Nym. Consp.
    var. oculatus *Boiss.*

121. D. microlepis *Boiss.* Boiss. Fl. Or.

122. D. Szowitzianus *Boiss.* Boiss. Fl. Or.

123. D. puberulus *mihi.* Herb. Kew: coll. Haussknecht in Luristan. — Puberulus 24 centim. Caules ramosi teretes 1-2-florales. Folia 25 mm., lanceolato-linearia acuta carinata adpressa; vaginâ purpureâ caulis diam. 2-plo longiore. Bracteæ 8 acuminatæ ovatæ, adpressæ. Calyx purpureus dentibus lanceolatis acuminatis. Petala contigua.

Subsect. 2. HEMISYRHIX.—Bracteæ atting. $\frac{1}{2}$ calycis longitudinem.

124. D. deltoides *L.* Nym. Consp.
    var. genuinus *L.*
    var. glaucus *L.*
    var. microlepis *Boiss.*

125. D. multisquamatus *mihi.* Kurdistan, Haussknecht, 1876; in Hb. Mus. Brit. et magnopere differt sp. 123. — Glaber 52 centim. Caules ramosi teretes. Folia linearia obtusa plana stricta adpressa; vaginâ flavâ caulis diam. æquante. Bracteæ 10 obovatæ mucronatæ purpureæ, adpressæ. Calyx apice attenuato dentibus lanceolatis acuminatis purpureis.

126. D. gaditanus *Boiss.* Nym. Consp. (Cadiz).
    var. roseo-luteus *Velen.* Œsterr. Bot. Zeit. 1887.

127. D. pubescens *S. et S.* Nym. Consp.

128. D. hypochloros *B. et Hldr.* Boiss. Fl. Or.

129. D. aridus *Griseb.* Nym. Consp.

130. D. alpinus *L.* Nym. Consp.
    var. pavonius *Tsh.*
    var. Semenovii *Reg. et Herd.*
    var. genuinus (*sec. Rchb.*).

131. D. brevicaulis *Fzl.* Boiss. Fl. Or.

132. D. versicolor *Fsch.* Ledebour, Flora Rossica.

133. D. diffusus *S. et S.* Nym. Consp.
    var. cylleneus *B. et Hldr.*

134. D. masmenæus *Boiss.* Boiss. Fl. Or.
    var. glabrescens *Boiss.*
    var. *K*arami *Bl.*—Folia flaccidiora. Bracteæ semi-herbaceæ
        margine rubellæ. Calyx viridis dentibus sanguineis.
    var. œtæus *Hldr.*—Græcia septent.
135. D. myrtinervius *Griseb.* Nym. Consp.
    var. deltoides *Griseb.*
    var. subalpestris *Hpe.*
136. D. campestris *M. B.* Nym. Consp.
137. D. aristatus *Boiss.* Boiss. Fl. Or.
    var. minor *Boiss.*
138. D. humilis *W.* Nym. Consp.
139. D. Buergeri *Miq.* Flora Japoniæ.
140. D. nitidus *W. et K.* Nym. Consp.
    var. obtusus *mihi.* — Folia latiora obtusiuscula ; prob. forma
        orientalis.—Jebel Muneitsi, Syria, 1879 ; Herb. Kew.
141. D. Seidlitzii *Boiss.* Boiss. Fl. Or.
142. D. elatus *Led.* Flora Altaica.
143. D. callizonus *S. et K.* Nym. Consp.
    var. Brandzæ *Panc.* Sehed. Fl. Austro-Hung.

Subsect. 3. Longisquamea. — Bracteæ calycem æquantes longitudine.

144. D. pruinosus *B. et O.* Boiss. Fl. Or.
145. D. pratensis *M. B.* Flora Taurico-Caucasica.
    var. guttatus *M. B.*—Cæspitosus. Bracteæ ovatæ.
146. D. suaveolens *Spreng.* Ledebour, Flora Rossica.
147. D. pallidiflorus *Ser.* Nym. Consp.
    var. ramosissimus *Pall.*
148. D. gelidus *Schott.* Nym. Consp.
    var. alpinus *Vill.*
149. D. glacialis *Hke.* Nym. Consp.
    var. neglectus *Lois.*
    var. subalpinus *Gaud.*

Sectio III. Caryophyllum.— Caules glabri. Bracteæ adpressæ. Petala dentata, imberbia. Calyx dentibus lanceolatis. Capsula ovoidea v. oblonga, nunquam cylindrica. Semina peltata.

Subsect. 1. Caryophylloides.—Species typicæ. Folia patentia. Calyx dentibus acuminatis. Capsula ovoidea.

150. D. Caryophyllus *L.* Nym. Consp. — Formarum omnium hic
    typus generis.
    var. genuinus *L.*
    var. acinifolius *Schur.*
    var. binatus *Schur.*
    var. Scheuchzeri *Jord.*
    var. collivagus *Jord.*
    var. carduinus *Ser.*
    var. divaricatus *d'Urv.*

151. . Henteri *Heuff*. Nym. Consp.
152. . caryophylloides *Schult.* Nym. Consp.
153. D. longicaulis *Ten.* Nym. Consp.
154. . Boissieri *Willk.* Nym. Consp.
155. . multinervis *Vis.* Nym. Consp.
156. . Arrostii P*resl.* Nym. Consp.
    var. uniflorus P*resl.*
    var. biflorus P*resl.*
157. D. Falconeri *Edgw.* Hooker, Fl. Brit. India, i.
158. D. crenatus *Thunb.* Flora Capensis, i.
159. D. sinensis *L.* Linn. Sp. Plant. ed. 1, p. 411.
    var. minor = D. Seguieri *Vill.* + D. caucasicus *M. B.*
    var. sylvaticus *Koch.*
    var. asper *W.*

Subsect. 2. SYLVESTRES.—Caules tenues. Bracteæ mucronatæ.
Capsula oblonga.

160. D. sylvestris. Nym. Consp.
    varr. prox. D. sylvestris *Wulf.*

a. sylvestris *Wulf.*	*e.* saxicola *Jord.*
b. brachycalyx *Huet.*	*f.* juratensis *Jord.*
c. pratensis *Jord.*	*g.* bracteatus *W. et L.*
d. binatus *Bartl.*	*h.* ebracteatus *W. et L.*

    varr. prox. D. sylvestris *Jacq.*

i. sylvestris *Jacq.*	*m.* consimilis *Jord.*
k. Bauhinianus *Noë.*	*n.* frigidus *Kit.*
l. orophilus *Jord.*	

161. D. laricifolius *B. et Reut.* Nym. Consp.
162. D. serratifolius *S. et S.* Nym. Consp.
    var. nazarens *Clk.*—Caules stricte ramosi.
163. . virgineus *Jacq.* Nym. Consp.
164. D. Balansæ *Boiss.* Boiss. Fl. Or.
165. . xylorrhizus *B. et Huet.* Boiss. Fl. Or.
166. . attenuatus *Sm.* Nym. Consp.
167. . sabuletorum *Willk.* Willk. et Lange, Prod. Fl. Hispan.
    var. pyrenaicus *W. et L.*
168. D. furcatus *Balb.* Nym. Consp.
169. D. siculus *Presl.* Nym. Consp.
    var. miniatus *Huet.*
170. D. cachemiricus *Edgw.* Hooker, Fl. Brit. India, i.
171. D. longiglumis *Del.* Richard, Tent. Fl. Abyssinicæ.—Glaucus,
    60 centim. Folia lineari-lanceolata plana; radicalia
    36 mm., caulinia 30 mm.; vaginâ caulis diam. æquante.
    Bracteæ 4, lanceolatæ ad calycis apices. Calyx dentibus
    lanceolatis. Petala contigua, lamina obovato-cuneata.
172. D. papillosus *Vis. et Panc.* Nym. Consp.
173. D. Jacquemontii *Edgw.* Hooker, Fl. Brit. India, i.

Sectio IV. Imparjugum.—Bracteæ nunquam 4. Petala dentata, v. integra, imberbia. Capsula cylindrica.

174. D. sulcatus *Boiss.* Boiss. Fl. Or.
175. D. syriacus *mihi.* Herb. Mus. Brit., Aucher-Eloy, no. 501 ; Syria bor. — Glaber. Caules ramosi teretes. Folia linearia patentia, radicalia 24 mm., obtusa carinata, caulinia 8 mm., acuta plana; vaginâ caulis diam. æquante. Flores paniculis laxis. Bracteæ 2 obovatæ, mucronatæ ad dimidium calycem patentes. Calyx apice attenuato dentibus lanceolatis acuminatis.
176. D. Gasparinii *Gus.* Nym. Consp.
177. D. Kremeri *B. et Reut.* Mumby, Enum. Pl. Algeriæ.
178. D. ciliatus *Gus.* Nym. Consp.
    var. racemosus *Vis.*
    var. cymosus *Vis.*
    var. Brocchianus *Vis.*
179. D. aragonensis *Timb.* Willk. et Lange, Prod. Fl. Hispan.
180. D. multiceps *Cst.* Nym. Consp.
181. D. Costæ *W. et L.* Willk. et Lange, Prod. Fl. Hispan.
182. D. Talyschensis *B. et Bu.* Boiss. Fl. Or.
183. D. stenocephalus *Boiss.* Boiss. Fl. Or.
184. D. repens *W.* Ledebour, Fl. Rossica.
185. D. fragrans *M. B.* Boiss. Fl. Or.
186. D. Legionensis *W. et L.* Willk. et Lange, Prod. Fl. Hispan.
187. D. holopetalus *Türcz.* Flora Capensis, i.
188. D. angolensis *Hiern ms.* Prope Caconda sub lat. 14° mer. Jul. 1880.—Glaber. Caules 30 centim., teretes, ramosi, tenues. Folia lanceolato-linearia adpressa ; radicalia 60–65 mm., 5-nervia, caulinia 50–55 mm., 3-nervia. Flores racemis laxis nec sæpius subsolitarii. Bracteæ 6–8 ovatæ ad dimidium calycem acuminatæ, adpressæ. Calyx dentibus lanceolatis acutis. Lamina dentata.

Sectio V. Tetralepides Leiopetala.—Bracteæ semper 4. Petala integra v. dentata, imberbia. Capsula cylindrica.

Subsect. 1. Hispanioides. — Caules ramosi. Bracteæ atting. $\frac{1}{3}$ calycis longitudinem.

189. D. hispanicus *Asso.* Nym. Consp.
    var. australis *Willk.*
    var. borealis *Willk.*
190. D. hirtus *Vill.* Nym. Consp.
    var. vivariensis *Jord.*
191. D. Requiennii *G. et G.* Nym. Consp.
192. D. cognobilis *Timb.* Nym. Consp.
193. D. albens *E. M.* Flora Capensis, i.
194. D. tripunctatus *S. et S.* Nym. Consp.
    var. Barati *Dur.*

Subsect. 2. Sætabensis.— Caules ramosi.   Bracteæ atting. $\frac{1}{2}$–$\frac{2}{3}$ calycis longitudinem.   Glabri.

195. D. Kamisbergensis Sond.   Flora Capensis, i. — Caules 37 centim., paniculati teretes.   Folia stricta, radicalia 42 mm., linearia acuta, caulinia 6–8 mm., elongato-linearia acuminata ; vaginâ caulis diam. æquante. Bracteæ obovatæ mucronatæ adpressæ.   Lamina integra rosea.
196. D. Planellæ Willk.   Nym. Consp.
197. D. Andersonii mihi.   Hb. Mus. Brit.   Syria, coll. Kotschy, 1855. — Caules paniculati teretes.   Folia linearia acuta adpressa ; vaginâ flavâ caulis diam. æquante.   Flores parvi geminati rosei.   Bracteæ adpressæ stramineæ. Calyx dentibus lanceolatis acuminatis purpureis.   Lamina obovato-cuneata.
198. D. Sætabensis Rouy.   Bull. Soc. Bot. France, 1881.— Cæspi-tosus.   Caules 40 centim., teretes.   Folia linearia, flaccida 3-nervia plana ; vaginâ caulis diam. æquante. Bracteæ ovali-lanceolatæ, mucronulatæ adpressæ.   Calyx apice attenuato dentibus lanceolatis acuminatis.   Lamina obovato-cuneata $=$ $\frac{1}{2}$ unguis.
 var. minor Rouy.
 var. media Bouy.
199. D. auraniticus Post.   Journ. Linn. Soc. vol. xxiv. p. 422.— Glaucus 30 centim.   Folia 40–45 mm., linearia acuta, canaliculata marginaliter pallida ; vaginâ caulis diam. æquante.   Bracteæ oblongo-lanceolatæ acutæ patentes. Calyx dentibus mucronatis.   Lamina obovato-spathulata.

Subsect. 3. Cintrani.—Caules simplices.   Bracteæ mucronatæ.

200. D. elongatus C. M.   Boiss. Fl. Or.
201. D. strictus S. et S.   Nym. Consp.
 var. grandiflorus Vis.⎫
 var. bebius Vis. ⎬Dalmaticæ.
 var. integer Vis. ⎭
202. D. procumbens Vent.   Boiss. Fl. Or.
203. D. leucophæus S. et S.   Nym. Consp.
 var. macropetalon Clem. — Petala suprà purpurea infrà pallidiora, latiora.
204. D. virgineus G. et G.   Nym. Consp.
 var. mauritanicus Ball.—Folia multum contorta.
205. D. micranthus B. et Hldr.   Boiss. Fl. Or.
 var. minor Boiss.—Petala minuta.
206. D. Haussknechtii Boiss.   Boiss. Fl. Or.
207. D. anatolicus Boiss.   Boiss. Fl. Or.
 var. parviflorus Boiss.
208. D. Kotschyanus Boiss.   Boiss. Fl. Or.
209. D. cintranus B. et Leut.   Nym. Consp.
210. D. insignitus Timb.   Nym. Consp.

211. D. algetamus *Grlls.* Herb. ipsius in Hb. Mus. Brit.—Cæspitosus, glaucus, glaber 22 centim. Caules teretes tenues.
Folia 8–9 mm., linearia acuta stricta 3-nervia carinata
adpressa. Flores geminati rosei. Bracteæ ovato-lanceo-
latæ ad ⅓ calycis partem, patentes. Calyx purpureus
apice attenuato. Lamina dentata. Hispania.
212. D. brachyanthus Boiss. Nym. Consp.
    var. montanus *Willk.*
    var. ruscinonensis *Willk.*
    var. humilis *Nym.*
    var. alpinus *W. et L.*
    var. nivalis *W. et L.*
213. D. Langeanus *Willk.* Willk. et Lange, Prod. Fl. Hisp. iii.

Subsect. 4. PUNGENTES.—Caules simplices. Bracteæ acuminatæ.

214. D. graniticus *Jord.* Nym. Consp.
215. D. leptoloma *Steud.* Steud. in Pl. Schimp. Abyssin. iii. (1761).
216. D. pungens G. *et* G. Flore Française (*non* Linn. Sp. Plant.).
217. D. serratus *Lapeyr.* Abr. Pyren. 241. — Caules 4-angulares.
Folia acuminata, elongato-linearia, 3-nervia, stricta,
radicalia 24 mm., caulinia 30 mm. Flores geminati,
rosei. Bracteæ lanceolatæ ad dimidium calycem patentes.
Calyx rubellus. Petala non contigua, lamina dentata.
218. D. acuminatus *mihi* (nomen vetus revictum). Syria. —
Cæspitosus, glaber. Caules teretes. Folia elongato-
linearia acuminata patentia 5-nervia, inferiora recurva,
superiora incurva; vaginâ folii diam. æquante. Bracteæ
lanceolatæ ad dimidium calycem. Calyx dentibus lanceolatis.—Hb. Mus. Brit., Aucher-Eloy, no. 526.
219. D. sphacioticus *B. et Hld.* Nym. Consp.
220. D. lactiflorus *Fzl.* Boiss. Fl. Or.
221. D. judaicus *Boiss.* Boiss. Fl. Or.
222. D. benearnensis *Loret.* Nym. Consp.
223. D. liboschitzianus *Ser.* Boiss. Fl. Or.
    var. integerrimus *Bge.* — Folia latiora. Bracteæ in aristam
longiorem calycem subæquantem. Lamina obovato-
cuneata, integerrima.
    var. multicaulis *B. et Hldr.*

Subsect. 5. GYMNOCALYX. — Caules ramosi. Bracteæ minutæ
scariosæ, adpressæ, arista incurrente ⅕ calycis.

224. D. cinnamomeus *S. et S.* Nym. Consp.
    var. pallens *S. et S.*
225. D. leptopetalus *W.* Nym. Consp.
226. D. subacaulis *Vill.* Nym. Consp.—Cæspitosus, glaber. Caules
8 centim., furcati, 4-angulares. Folia 22–24 mm. patentia
canaliculata; vagina caulis diam. æquante. Calyx purpureus apice attenuato dentibus lanceolatis mucronatis.
Petala purpurea contigua, lamina integra obovato-
cuneata, = unguis.

Subgenus III. PROLIFERASTRUM.—Herbæ annuæ. Folia bracteiformia (suprema) sub floribus densa. Flores capitati. Bracteæ 2. Calyx 15-costatus, apice pentagono attenuato. Petala retusa. Torus parvus. Capsula oblonga.

227. D. Cyri *F. et M.* Ind. iv. Hort. Petrop. p. 34.
228. D. glumaceus *B. et C.* Boiss. Fl. Or.
229. D. obcordatus *R. et M.* Boiss. Fl. Or.
230. D. prolifer *L.* Nym. Consp.
      var. scabrifolius *Clav.*
      var. lævis *Clav.*
231. D. velutinus *Gus.* Nym. Consp.
      var. Sartorii *Frhl.* Ellwang, Sartori 240.

The investigation of a natural genus and of the specific forms which it includes, and the close study of the special morphology of such a genus, often throw a light on the physiological significance of a much larger section than that which comprises the forms under observation. It is for this reason, therefore, that it is interesting to note the various forms, referable to a natural genus, existing under various conditions of climate and locality, and subjected to the diverse circumstances of their environment. As soon as we begin to group morphological facts into general propositions, we cannot fail to recognise *a posteriori* what was to be expected *a priori*, the indefiniteness of distinction. And though the general phases of organisation in allied species may appear to mark off one from another, it is impossible to separate absolutely groups of characters which structural affinity links. It is by these means that we are enabled to see the gradual differentiation of allied forms from a few or one primary form after a succession of generations under favourable conditions or otherwise for the modifications of certain organs: for Evolution, whether referable to the vegetal, animal, or social organism, implies insensible modifications and gradual transitions.

Dianthus is a genus in which very many species can most readily be crossed, and which lend themselves easily to variation and hybridity; though the correspondence between systematic affinity and the facility of crossing in other genera is by no means strict: for in the allied genus *Silene*, which contains a much larger number of species, and in which the systematic affinity is as well developed, the most persevering efforts have failed to produce between extremely close species a single hybrid. DeCandolle considered plants of the same species (1) to have in common numerous and important characters, which are permanent during several generations, under the influence of varied environments; (2) to have flowers easily fertilising among themselves, and producing seeds most usually fertile; (3) to be affected by temperature and other external agents in a similar or nearly similar manner; (4) in short, as plants of similar structure, known certainly to have descended from a common stock for many generations. These data lead up to Darwin's conclusion that botanical species are only a higher and more permanent class of varieties. The gradation of varieties by art, with consequent deterioration of organs essential to the natural maintenance of the species, seems to be done with equal efficacy, though more slowly, by nature, without such deterioration of vital parts, and adapted for the balance of their internal relationships with those of the environment; and resulting in the formation of varieties, which, fertilizing among themselves, and remaining permanent through successive generations, ultimately attain to specific rank.

In this genus the less constant characters, and for this purpose more valuable for the better diagnosis and distinguishing marks of the species are—the number and form of the bracts forming the epicalyx, the form of the lamina of the petals and their apposition, the character of the calyx-teeth, the form and structure of the capsule, the form and structure of the seeds, and the disposition of the fascicles of veins in the leaves of the barren shoots and flowering stems. Characters more constant are—the form of the calyx, the nature of the margin of the lamina of the petals, presence or absence of a beard at the junction of the lamina with the unguis, the relative lengths of the petals, filaments, and styles of the flower, the breadth of the vagina of the leaf, and the disposition of the flowers.

From the materials accumulated for a monograph on the genus I have made this abstract in advance, and hope to work out its completion shortly. Great difficulty has been experienced in determining a *via media* in the matter of species; in avoiding the tendency to species-splitting to which the systematist is liable, as well as the habit of lumping allied forms more congenial to the synthetical botanist. The wholesale exclusion of synonyms far exceeding the number of specific forms has itself entailed much research. For the purposes of its study I have made use of all the available material in the Herbarium of the Museum of Natural History at South Kensington, in the Kew collections, in the National Herbarium at the Jardin de Plantes in Paris, and in the Botanic Garden of Brussels; as well as from many recent specimens kindly sent by botanists. I have especially to thank the late Prof. Josef Pancic, of Belgrade, for a valuable collection of specimens from Servia, Bulgaria, and Roumania.

And here I am led to believe, as many others have found, that the deeper one pursues researches into any single small group of animals or plants, the more one is impressed with the consistent phenomena of variation, even when carried into the minor details of natural processes. As Herbert Spencer says, these phenomena are unobtrusive while the fairly-uniform conditions of a species maintain fair uniformity among the physiological units of its members; but they become obtrusive when differences of conditions, entailing considerable functional differences, have entailed decided differences among these units; and when the different physiological units, differently mingled in every individual, come to be variously segregated and variously combined.

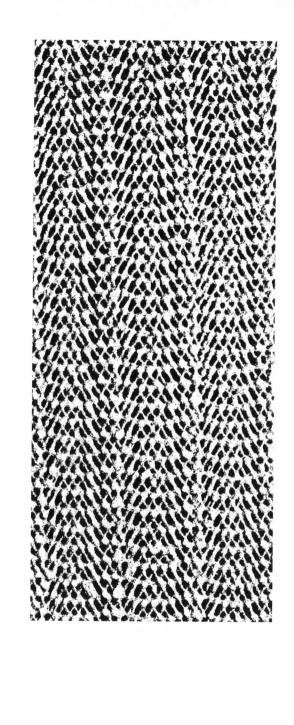

CPSIA information can be obtained
at www.ICGtesting.com
Printed in the USA
BVHW051107250219
541084BV00011B/1443/P